IMAGES
of America

NEW ORLEANS TELEVISION

Two of the biggest local television stars in the 1960s, Dr. Momus Alexander Morgus (left) and Chopsley, his faithful hulk of an assistant, are seen at work inside their lab during a WWL-TV broadcast. (Tom Bourgeois collection.)

On the Cover: Terry Flettrich, as local children's television icon Mrs. Muffin, serves birthday cake to some of the youngsters lucky enough to appear on her hugely popular 1950s WDSU program. (Paul Yacich collection.)

IMAGES
of America

NEW ORLEANS TELEVISION

Dominic Massa

ARCADIA
PUBLISHING

ISBN 978-0-7385-5404-4

Published by Arcadia Publishing
Charleston SC, Chicago IL, Portsmouth NH, San Francisco CA

Printed in the United States of America

Library of Congress Catalog Card Number: 2008920488

For all general information contact Arcadia Publishing at:
Telephone 843-853-2070
Fax 843-853-0044
E-mail sales@arcadiapublishing.com
For customer service and orders:
Toll-Free 1-888-313-2665

Visit us on the Internet at www.arcadiapublishing.com

A vintage view of the WWL studios during a newscast in the 1960s shows news anchor Jim Kincaid (left), sports anchor Hap Glaudi (center), and weathercaster Don Westbrook (right) seated behind the anchor desk. (Courtesy of WWL-TV.)

CONTENTS

Acknowledgments

While television is a visual medium, it is also a fleeting one, and so many of the memories of local television's golden age are just that—memories, without any photographs to help us remember. Thankfully, we do have dozens to share because of the efforts of pioneers who took it upon themselves to preserve the history.

At the top of the list is Paul Yacich, who was with WDSU-TV from the beginning. My deepest thanks go to him for sharing his many photographs and memories, as well as his friendship. I am also indebted to Joe Bergeron of Bergeron Studio and Gallery for sharing his vast collection of photographs.

Among the many others who shared photographs are Joe Budde, Bob and Jan Carr, Suzi Fontaine, Cathy Jacob Gaffney, Phil and Freida Johnson, John Pela, Terry Flettrich Rohe, Al Shea, and Don Westbrook.

I must thank my family, in particular my parents, Diane and Joe Massa, whose loving support means everything.

Beth Arroyo Utterback, Peggy Scott Laborde, and Errol Laborde, who helped launch my television career as a WYES-TV intern, have remained dear friends. As with everything I do, they have given important guidance and encouragement.

At WWL-TV, my thanks go to Bud Brown and Chris Slaughter for their support. Also special thanks to Jimmie Phillips, Sandy Breland, and Mark Swinney, as well as to Angela Hill and J. Michael Early, two beloved figures whose talents have made WWL the amazing television station it is today.

Even though I work for their competition, the management at WDSU-TV has been supportive. Thanks to Joel Vilmenay and Mason Granger.

Many talented colleagues and friends at WYES contributed to this project during the production of the documentaries *New Orleans TV: The Golden Age* and *Stay Tuned: New Orleans' Classic TV Commercials*. Thanks to Larry Roussarie, Maria D. Estevez, Hal Pluche, Kathy Burns, and Mark Larson for sharing their artistic talents. My appreciation also goes to general manager Randy Feldman.

I must thank the writers at the *Times-Picayune*, whose work on this topic has educated and inspired me over the years: Mark Lorando, Benjamin Morrison, Renee Peck, Dave Walker, and especially David Cuthbert.

Finally, thanks to the team at Arcadia Publishing and editor Katie Stephens, who helped turn a television producer's ideas into a book, something that sounds easier than it is.

INTRODUCTION

Do you remember when Terry Flettrich was Mrs. Muffin? How about when Mel Leavitt was Mr. Mardi Gras or when Nash Roberts was the Jax weatherman? Maybe you have better memories of Miss Ginny or Uncle Henry. How about Wayne Mack as the Great MacNutt? Or is it *Midday* you think of when someone says his name? What about Morgus the Magnificent or John Pela? If you grew up in New Orleans, or have spent any amount of time here, chances are that you answered yes to all of those questions.

More than 50 years ago, with the flip of a switch and the turn of a dial, local television became part of New Orleans culture. Ask local baby boomers to recall their favorite shows or stars, and the answers are much the same. There are unforgettable memories of growing up while watching this new medium develop. Sometimes those memories are of watching the test pattern—the only thing there was to watch at some points during the day. Still, it was new, it was exciting, and it was New Orleans.

"Everything we did in those early days was new," explains WDSU founding father Edgar Stern Jr. "Everything was a first, everything was an experiment, which made it a lot of fun."

It was also unpredictable. Two months after New Orleans' first television station signed on the air, its most popular program was "To Be Announced." At least, those are the words that popped up more than any other title on the WDSU program schedule in February 1949. Fortunately, there were only a few dozen television sets in the city at the time. By the end of that same year, the number of New Orleanians with a television set grew to 15,000. One year later, it was closer to 40,000.

When asked about their role, most of the television pioneers will modestly say they were simply in the right place at the right time, but in reality, it is what they did with the new medium that sets them apart. Many of them had a background in radio and tried to transfer their talents to this new thing called television.

There was Terry Flettrich, the radio host who was recruited by WDSU to help "sell" New Orleanians on the idea of television in the months before the station signed on the air. She produced and hosted demonstration shows from the D. H. Holmes department store cafeteria, which had been converted into a studio. As the first lady of New Orleans Television, she went on to contribute some of the medium's most memorable moments, most notably as Mrs. Muffin and the star of *Midday*, one of the most popular programs in local history.

Mel Leavitt, the dashing young sports announcer, came to New Orleans from New York for a job at Channel 6 in 1949 and never left. A true renaissance man, Leavitt anchored sportscasts, hosted talk shows, produced documentaries, anchored Carnival coverage, and so much more.

"Mel Leavitt is and will always be Mr. Television in New Orleans," says former WDSU colleague Al Shea, whose Channel 6 career began in the mid-1950s.

Creativity, originality, and personality were hallmarks of the golden age of television in New Orleans. How else do you explain Morgus the Magnificent or the Great MacNutt? What else could account for the sportscasting success of someone like Hap Glaudi? Where else could a young comedian named Dick Van Dyke sharpen his talents, in hopes of one day landing his own network show?

Local television has also been about longevity. Meteorologist Nash Roberts spent 50 years on the air here, becoming the most trusted man in New Orleans, at least when it came to hurricanes. Alec Gifford's 50 years in local television included coverage of every major news story imaginable. Phil Johnson bid viewers "Good evening" for 37 years as Channel 4's bearded bard, delivering nightly editorials. His colleague Don Westbrook spent nearly 40 years in front of the WWL weather map.

Add to that roster a cute married couple named Bob and Jan Carr, a dance show host named John Pela, a tough investigative reporter named Bill Elder, and a gifted writer named Jim Metcalf, and one starts to see why it truly was a golden age of television—for the people in front of and behind the camera and for those watching at home.

This is what Channel 4 news looked like in the 1960s in New Orleans—weather from Don Westbrook, sports with Hap Glaudi, and the news from anchor Bob Jones. (Courtesy of WWL-TV.)

One

ON THE AIR

The year is 1948, and in New Orleans, many people haven't even seen a television set, much less owned one. Even if they have, there aren't many programs to tune in to—for now. That all changes when WDSU signs on the air December 18, 1948, with a live broadcast from the Municipal Auditorium. WDSU is the first television station in Louisiana, the sixth in the South, and the 48th in the United States.

Channel 6 had somewhat humble beginnings but deep pockets, since its owner, Edgar Stern Jr., came from a wealthy family who was determined to help launch the young man on a career. Stern and chief engineer Lindsey Riddle would lead the team that helped build the station from scratch. Most of them had a background in radio but little experience in television.

The early crew helped put an antenna atop the Hibernia Bank, then the tallest building in town, and set up studios there and at the D. H. Holmes and Werlein's department stores to help sell New Orleanians on the new technology.

One of the early staffers who helped produce and host those first demonstration programs was a local radio star named Terry Flettrich. She remembers running around D. H. Holmes wearing black, white, and gray makeup, at the time believed to be preferable for performers who would be seen on the small black-and-white television sets.

In its early years, the station would help build the careers of television icons named Mel Leavitt, Nash Roberts, Bill Monroe, Alec Gifford, and even a future comedy star named Dick Van Dyke.

During its first nine years on the air, Channel 6 was the only station in town. For that reason, it was able to build a bond with viewers that would be hard to break. But the Jesuits of Loyola University, who had pioneered radio broadcasting in Louisiana with WWL in 1922, were willing to try. In 1957, Loyola introduced WWL-TV Channel 4. Like their WDSU counterparts, many of the early WWL stars, including Jill Jackson and Henry Dupre, had built their careers in radio, which eased their transition to the small screen.

When WDSU signed on the air in 1948, its antenna was placed atop the tallest building in town, the Hibernia Bank building on Carondelet Street, which also housed the station's control room, studios, and offices. (Paul Yacich collection.)

WDSU owner Edgar Stern Jr. was guided into the broadcasting business by his father, Edgar Bloom Stern, and Lester Kabacoff, the elder Stern's attorney and executive assistant. The two men helped nudge Stern into television after he returned home from World War II, where he was a radar engineer. (Author's collection.)

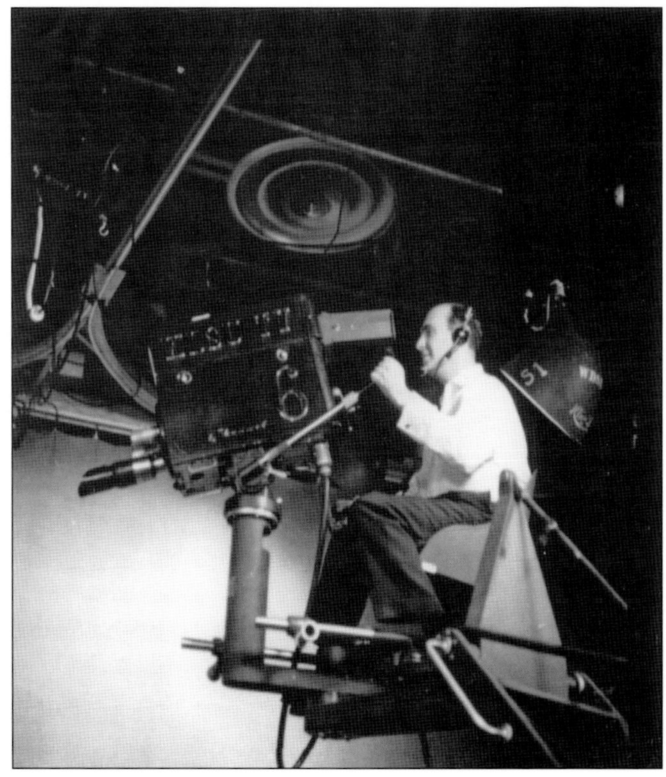

Edgar Stern Jr.'s fascination with the technical side of television helped WDSU earn its position as the city's preeminent station in the 1950s and 1960s. Stern, shown here behind a WDSU camera, helped build the station's reputation for quality programming and news coverage. (Paul Yacich collection.)

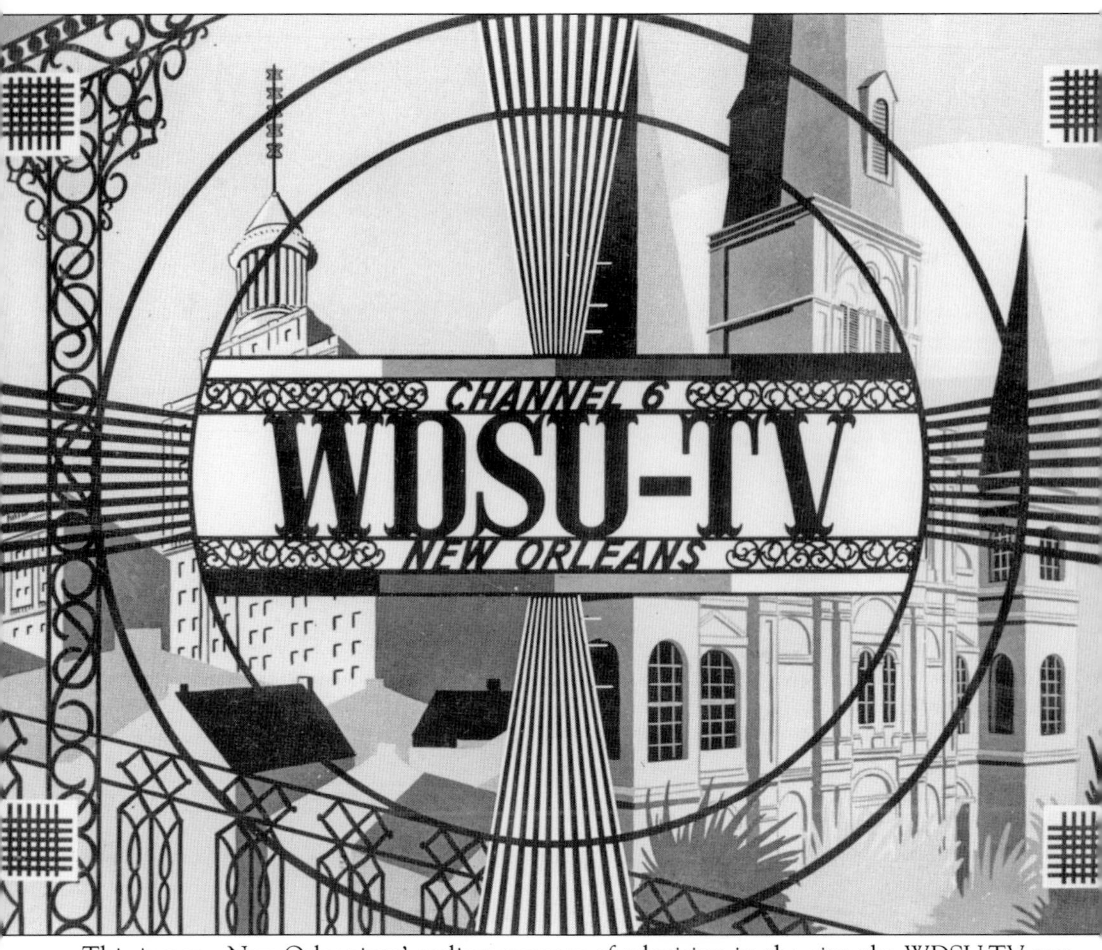

This is many New Orleanians' earliest memory of television in the city: the WDSU-TV test pattern. For the first few months the station was on the air, this French Quarter balcony scene was transmitted for most of the morning to allow viewers and technicians installing new sets across the city to make adjustments to their picture. (Paul Yacich collection.)

WDSU chief engineer Lindsey Riddle was with the station from the beginning, helping to lead the team that installed and tested the technology needed to begin broadcasting in 1948. (Paul Yacich collection.)

Paul Yacich began his association with WDSU in 1947 as an engineer with WDSU radio. He helped build the new WDSU-TV facility in 1948 and later became a member of the production department as an award-winning producer and director. (Paul Yacich collection.)

In the months leading up to the first WDSU broadcast, Terry Flettrich produced daily fashion, makeup, and cooking shows from the D. H. Holmes cafeteria. The demo programs aired on display television sets scattered through the department store. (Author's collection.)

Naomi "Nonie" Bryant (right) was one of the early stars of WDSU, well known for her singing and performing abilities. Bryant, shown here with a costar inside the small Hibernia Bank studios, later became the wife of "Mr. Television," Mel Leavitt. (Paul Yacich collection.)

An early WDSU studio production features one of the city's earliest television chefs, Lena Richard (center), who hosted her own cooking show on Channel 6. Cohosting is announcer Woody Leafer (right), and operating the camera is Ken Muller, who would later become a WDSU director. (Daphne Muller Eyrich collection.)

Some of the first remote television programs in New Orleans originated from the Marine Room of Lenfant's Restaurant on Canal Boulevard, where WDSU set up cameras to film the performance of the Basin Street Six, which featured a young Pete Fountain on clarinet (stage right). (Paul Yacich collection.)

A native of St. Louis, Missouri, Mel Leavitt was one of the early stars of WDSU, joining the station as a sports announcer and special events director in 1949. Leavitt later branched out into news and features, hosting a wide variety of programs, anchoring Carnival coverage, and writing documentaries. (Paul Yacich collection.)

By 1950, WDSU had outgrown its downtown digs and settled into new headquarters in the French Quarter at 520 Royal Street, inside the historic Brulatour mansion. Iron gates opened onto one of the most recognizable courtyards in the city. Few stations in the country can claim this kind of real estate in their history, but WDSU was a standout from the start. (Author's collection.)

Camera operators George Cuccia (left) and Buddy Rizzuto (far right) focus on, from left to right, producer/director Irwin Poche, host Mel Leavitt, and an unidentified guest during an early WDSU production in the studio on Royal Street. (Paul Yacich collection.)

WDSU served as the launching pad for the career of actor and comedian Dick Van Dyke, who came to Channel 6 from Atlanta in 1955. His stint in New Orleans as a staff announcer and host of his own daily show lasted only a matter of months before he took a job in New York with CBS. (C. F. Weber Collection, courtesy of Bergeron Studio and Gallery.)

During one of his WDSU programs, Dick Van Dyke adopts the pratfall pose for which he would later become famous on his groundbreaking CBS sitcom. (Paul Yacich collection.)

Dick Van Dyke and a costar appear on the set of his WDSU show. Van Dyke later recalled that his show featured comedy bits and music from a trio. (C. F. Weber Collection, courtesy of Bergeron Studio and Gallery.)

Gaines "Gay" Batson's voice probably helps transport viewers back to the earliest days of television in New Orleans. As WDSU's chief announcer, Batson's distinctive voice accompanied nearly every program or commercial in the formative years. (Author's collection.)

Pioneering WDSU staffers Woody Leafer (left) and Ray Rich work inside the station control room on Royal Street, a glassed-in booth that looked out onto the floor of the large studio below. (Paul Yacich collection.)

Announcer Dick Bruce established his career behind the microphone for WDSU radio and television before becoming much better known as the pitchman for McKenzie's bakery. (Author's collection.)

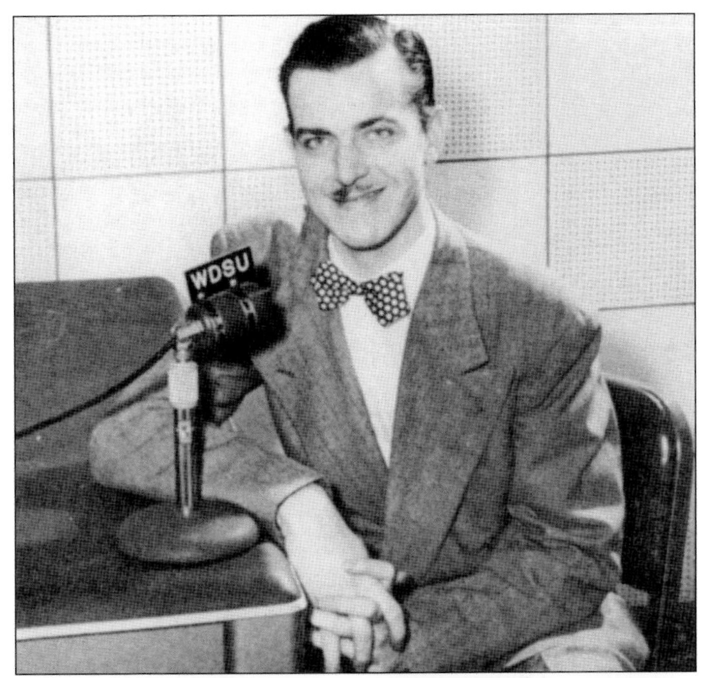

Ken Muller operates a WDSU camera during a remote broadcast at the Fair Grounds racecourse. An original WDSU staff member, Muller would later become a Channel 6 producer/director. (Daphne Muller Eyrich collection.)

One of the more unique WDSU productions involved the construction of a giant swimming pool inside the studio for a program on swimming and water safety. Note the glass window that was cut into the front of the pool so the cameras could see the action. (Paul Yacich collection.)

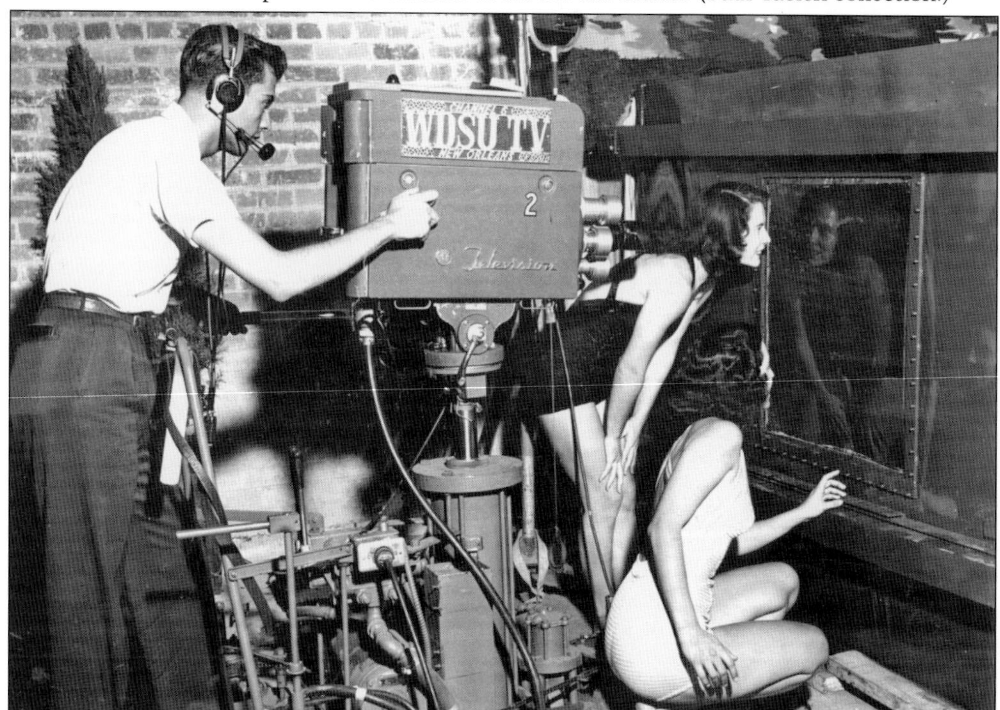

Paul Yacich (left) operates a camera during a WDSU studio production involving the station's 20,000-gallon swimming pool as two of the underwater actresses, Phyllis and Sydney Massicot, look on. (Paul Yacich collection.)

A. Louis Read had a successful career in sales and management at WWL radio before joining the staff at WDSU-TV three months after it signed on the air. The New Orleans native went on to serve as station general manager and executive vice president. (Paul Yacich collection.)

WDSU station manager Robert Swezey worked as vice president and general manager of the Mutual radio network before coming to New Orleans. Station owner Edgar Stern gives him credit for realizing the importance of television news in the early days and for hiring Bill Monroe as WDSU's first news director. (Paul Yacich collection.)

Mobile production units, like this converted Volkswagen van with a camera platform on the roof, would travel to outlying areas for WDSU's remote broadcasts. Here the van is shown driving through the French Quarter. (Paul Yacich collection.)

Merlin "Scoop" Kennedy (center) and Marie Matthews (left) appear on an early WDSU cooking program, along with announcer Jack Alexander (right). Matthews was a fixture in the kitchen at WDSU for more than 40 years, assisting on-air chefs on *Midday* and countless other productions. (Paul Yacich collection.)

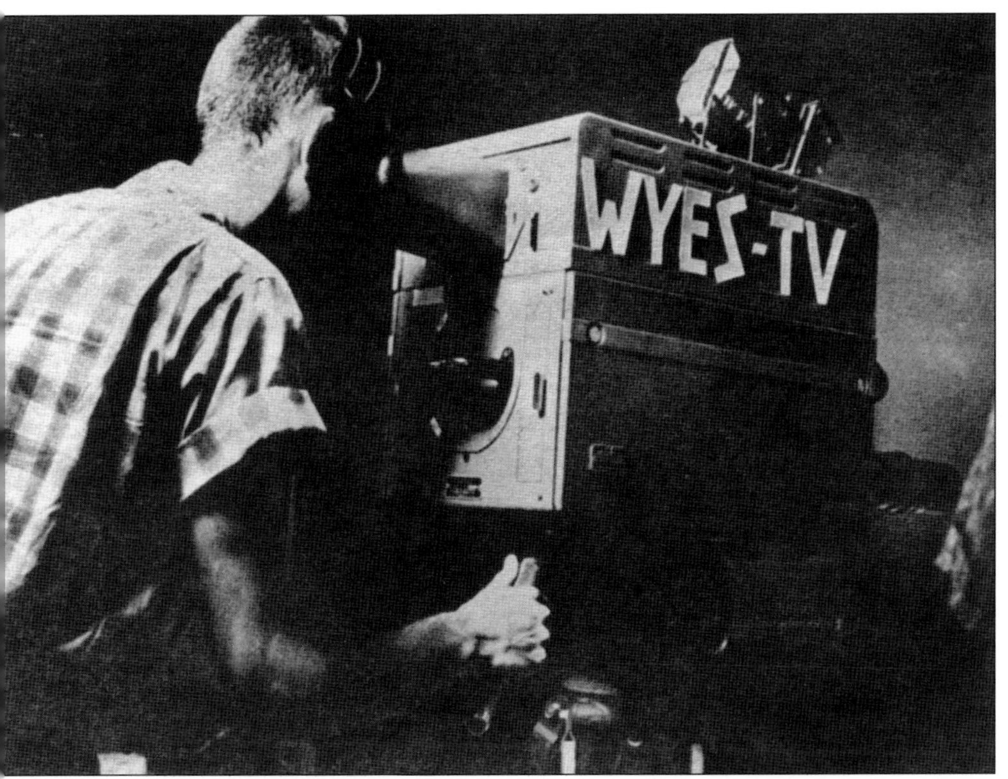

New Orleans' public television station WYES signed on the air April 1, 1957, first as Channel 8, then Channel 12. The station was the dream of a group of community activists convinced of the power of television as an educational tool. (Courtesy of WYES-TV.)

Members of the early WYES staff prepare for production inside the studios near City Park on Navarre Avenue. An early favorite, Marta Lamar's *Garden Show*, premiered in 1959. The station's first Showboat Auction fund-raiser was organized in 1967. (Courtesy of WYES-TV.)

WWL-TV billed itself as "the newest view" in town when it signed on the air September 7, 1957. Owned by the Jesuits of Loyola University, the television station built on the success of WWL radio, which signed on in 1922. (Author's collection.)

Members of the early WWL-TV staff pose for a photograph inside the North Rampart Street studio. From left to right are (first row) Juozas Bakshis; (second row) Bill Dean, Jill Jackson, and four unidentified; (third row) Rupert Copponex, Henry Turner, Al Moliere Jr., two unidentified, Mike O'Connor, Hal Cairns, Bob Lawton Jr., two unidentified, Al Shear, and Harold Hisey. (Courtesy of WWL-TV.)

This photograph, snapped by a viewer, shows how WWL's first night of broadcasting would have looked to those watching on September 7, 1957. Station program director Ed Hoerner helped emcee the inaugural broadcast. (Ed Hoerner Jr. collection.)

WWL radio veterans Henry Dupre (left) and Ed Hoerner (right) were among the hosts of *The Newest Look*, the program that introduced Channel 4 to New Orleans viewers. It was sponsored by D. H. Holmes and Regal Beer. (Courtesy of the Loyola University J. Edgar and Louise S. Monroe Library.)

Following the local premiere broadcast, featuring music from a studio orchestra, Channel 4's first network programs included *The Gale Storm Show*, *Gunsmoke*, and the Miss America pageant. (Courtesy of the Loyola University J. Edgar and Louise S. Monroe Library.)

New Orleans mayor deLesseps "Chep" Morrison appeared on WWL's premiere broadcast, congratulating Loyola University on the birth of its new station. (Courtesy of the Loyola University J. Edgar and Louise S. Monroe Library.)

For its studios, WWL-TV converted an old Zetz 7-UP bottling plant on the edge of the French Quarter. WWL-AM, which had its roots on the Loyola University campus before moving to the Roosevelt Hotel, would later be housed in the same Rampart Street complex. (Author's collection.)

This photograph shows an early WWL studio production. The early WWL schedule included both local programming and CBS network shows. (Author's collection.)

Engineer Francis Jacob Jr.'s connections to WWL dated back to his days as a student at Loyola University. He began working at Loyola's radio station in 1930, later helping to build WWL-TV, where he would serve as chief engineer. (Cathy Jacob Gaffney collection.)

WWL-TV chief engineer Francis Jacob Jr. poses for a picture while on a remote broadcast from the Pontchartrain Beach amusement park in the early 1960s. (Cathy Jacob Gaffney collection.)

Newspaper columnist and radio show host Jill Jackson (left) was one of the WWL radio stars who made the transition to television in 1957. Known as one of the first female sportscasters in America, Jackson also interviewed celebrities, like actor Gary Merrill. (Author's collection.)

Henry Dupre helps introduce viewers to Loyola University's new television station by encouraging them to appear on the air with him at a WWL remote broadcast from a local bank. (Author's collection.)

Channel 4 fixture Henry Dupre made the switch to television from a successful career in New Orleans radio, where he was well known as a star of the very popular *Dawnbusters* program on WWL. (Author's collection.)

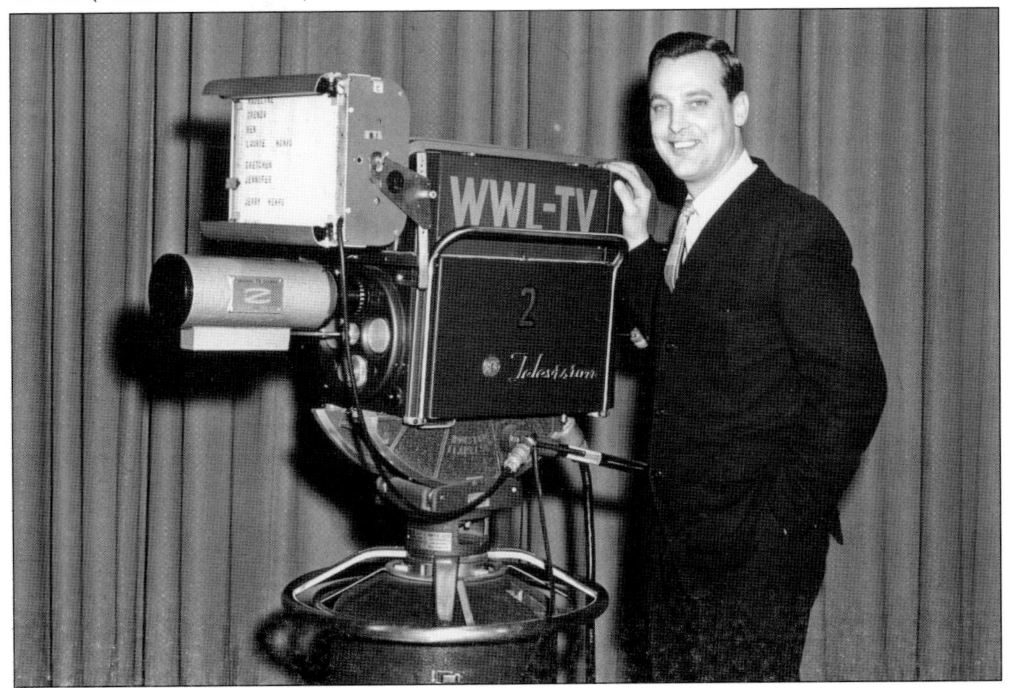

Early WWL-TV announcer and host Bill Wilson poses beside a vintage WWL camera in a 1960s publicity photograph. A native of West Virginia, Wilson worked at Channel 4 for 25 years. (Courtesy of WWL-TV.)

WWL-TV general manager J. Michael Early (right) accepts an Emmy award citation from Rosel Hyde (left), the chairman of the Federal Communications Commission. Early is credited with helping to establish WWL as the dominant television station in New Orleans soon after he took the reins of the station in 1961. (Courtesy of WWL-TV.)

Two

Mrs. Muffin, Uncle Henry, MacNutt, and More

Even though the medium was new and there were far fewer channels to watch during the golden age of New Orleans television, there was a wealth of local programming for kids.

One of the earliest stars was one of local television's first faces—Terry Flettrich, who starred on WDSU as the title character on *Mrs. Muffin*. She told stories, served cookies, and welcomed boys and girls to her quaint cottage, where a sign reminded young guests to "Be Pleasant."

On Channel 4, "Uncle Henry" Dupre welcomed children into WWL's studio each week for cartoons with *Popeye and Pals*.

New Orleans, like other cities, had a *Romper Room* schoolhouse, where Miss Ginny, and later Miss Linda, taught us the difference between a "do bee" and a "don't bee." As the lyrics to the familiar song explained, "I always do what's right, I never do anything wrong. I'm a *Romper Room* do bee, a do bee all day long." Earning a spot on *Romper Room* required writing a letter to the station. But through the magic of television, even those boys and girls who didn't appear on the show had a special connection, through the magic mirror into which the host looked to "see" children in the home audience.

WDSU's Wayne Mack hosted one of the most popular shows in the 1960s as *The Great MacNutt*. As a wacky film director, Mack would introduce Three Stooges shorts, tell jokes, and entertain his audience.

A long-running WDSU program, *Let's Tell a Story*, featured a huge storybook set, out of which an elf would appear and introduce the storyteller.

New Orleans icon Mr. Bingle made annual appearances on WDSU, coming to life at Christmas time to subtly help pitch products for the Maison Blanche store.

WDSU also had Captain Sam and Deputy Oops, while WWL persuaded announcer John Pela to star as Captain Mercury and Johnny Miller starred as the host of *Johnny's Follies* on WJMR, later WVUE-TV.

The shows were creative, imaginative, and, most of all, fun. For baby boomers in New Orleans, the memories remain vivid, even if most of what they saw was in black and white.

Terry Flettrich and friends appear on the set of one of her WDSU *Mrs. Muffin* programs. A 1951 *Item* newspaper article described Mrs. Muffin as living "in an old-fashioned cottage" where the activities ranged from "reading favorite fairy tales to showing children how to make puddings." (Paul Yacich collection.)

Mrs. Muffin helps celebrate the special birthdays of two young guests on the set of her program. (Terry Flettrich Rohe collection.)

Children were invited to write a letter to Mrs. Muffin, with the goal of being selected as one of the 50 or so children who appeared on the show. Every child received this postcard, which read, "Little friend, thank you for your very nice letter. I enjoyed it so much, and I hope I'll be hearing from you again soon." It was signed, "With love, Mrs. Muffin." (Sue "Suzi" Fontaine collection.)

A young *Alice in Wonderland* character joins *Mrs. Muffin*'s Terry Flettrich (right) on the set of her WDSU show, most likely to coincide with the 1951 release of Walt Disney's animated film *Alice in Wonderland*. (Sue "Suzi" Fontaine collection.)

Premiering in 1950, *Mrs. Muffin* had various time slots (including on Tuesday and Thursday afternoons as well as Saturday mornings) and various incarnations on WDSU, from *Mrs. Muffin's Birthday Party* to her *Surprise Party*, *Christmas Party*, and *Magic Cottage*. Whatever the name, the premise and personality were the same. (Paul Yacich collection.)

Generations of New Orleanians remember the huge storybook set of *Let's Tell A Story*, the popular story-time program that ran on WDSU-TV for decades. It was produced by the local chapter of the National Council of Jewish Women. (C. F. Weber Collection, courtesy of Bergeron Studio and Gallery.)

Let's Tell A Story had a remarkable run on WDSU, growing out of a similar Channel 6 program called *Magic Tree*, which premiered in 1951. The storytelling continued until 1974. (C. F. Weber Collection, courtesy of Bergeron Studio and Gallery.)

Al Shea starred as "Deputy Oops" on the WDSU children's program *Adventures in Fun*, sort of a spinoff from another children's program in which Shea starred, called *Tip Top Space Ship*. Shea replaced Ed Nelson on that program when Nelson left to pursue a successful career in Hollywood. (Al Shea collection.)

On *Adventures in Fun*, Al Shea's costar literally lit up the screen. To the right of the picture, Shea, as Deputy Oops, talks to a lightbulb on a stand, which he operated by pressing down a foot pedal. (Al Shea collection.)

Al Shea stars as Deputy Oops with two of his young costars on *Adventures in Fun*. (Joe Budde collection.)

This is a view inside WWL's *Romper Room* schoolhouse, with "Miss Ginny" Hostetler teaching her guests the difference between "do bees" and "don't bees." Third from left in this May 1963 photograph is future WWL-TV reporter Meg Farris. (Meg Farris collection.)

This photograph of "Miss Ginny" Hostetler and the kids in the *Romper Room* schoolhouse is from a WWL promotional piece. In it, the station boasts of Hostetler's popularity among children and adults, saying "The kids know her—thousands of them who have lovingly followed her gentle leadership in a streamlined television nursery school." (Phil Johnson collection.)

Ginny Hostetler is photographed outside of her *Romper Room* classroom in a WWL-TV promotional brochure. Like most of the early Channel 4 staff, Hostetler was versatile as a host and performer, and the station boasted she could "pitch" products for advertisers in commercials. (Phil Johnson collection.)

In its later incarnation on WVUE-TV, *Romper Room* was hosted by local actress "Miss Linda" Mintz, who was selected from among some 200 women who auditioned for the role. (Linda Mintz collection.)

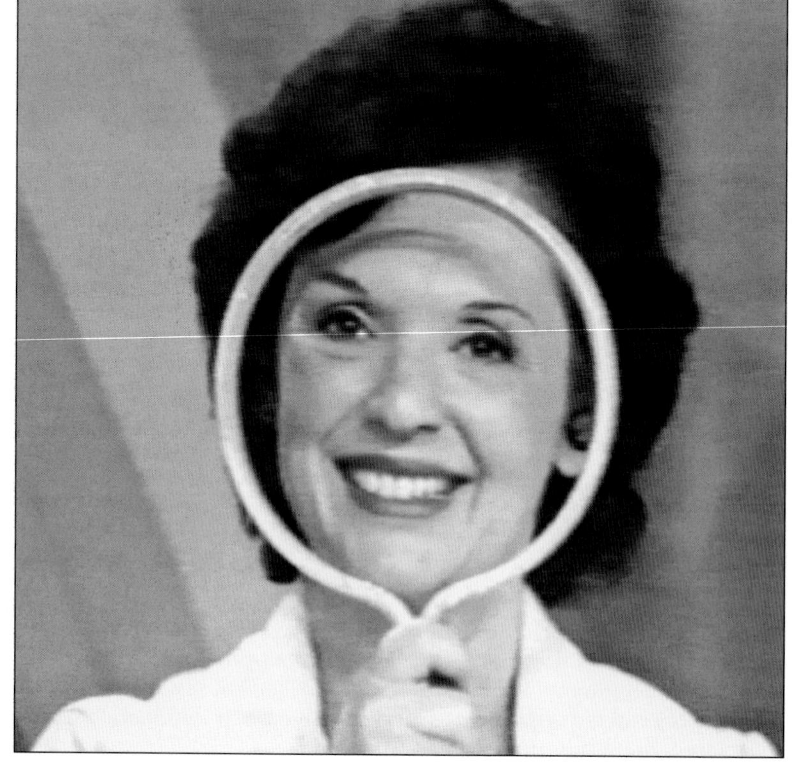

Did Miss Linda ever "see" you in her magic mirror? Remember how each *Romper Room* show ended? "Romper bomper, stomper boo. Tell me, tell me, tell me do. Magic mirror, tell me today, did my friends at home have fun at play?" (Linda Mintz collection.)

"Uncle Henry" Dupre was known to a generation of New Orleanians as a radio icon, but to their children and grandchildren, he will always be remembered for *Popeye and Pals*, the television show he hosted on WWL-TV in the 1950s and 1960s. (Courtesy of WWL-TV.)

A seasoned broadcaster, Henry Dupre was as comfortable on the set of his *Popeye and Pals* show as he was behind the microphone on the long-running *Dawnbusters* radio show on WWL. On *Popeye*, he and his gallery of guests introduced cartoon shorts featuring the animated sailor. (Tom Bourgeois collection.)

"Uncle Henry" Dupre, on the set of *Popeye and Pals*, poses before a huge cake bearing the Popeye character. The program was an early favorite on Channel 4, debuting the week after the station signed on the air in September 1957. (Author's collection.)

WWL staff announcer John Pela hosted *Popeye and Pals* for a short time following the retirement of "Uncle Henry" Dupre in the 1960s. (John Pela collection.)

Wayne Mack was wildly popular as the Great MacNutt on WDSU, a role he began soon after coming to New Orleans in 1958. An Illinois native, Mack came here to work as an announcer, later gaining recognition as a sidekick on *Midday* and as the station's sports director. (Mary Lou Mack collection.)

WDSU musical director Pete Laudeman, seated at the colorfully decorated piano, accompanies the Great MacNutt in this promotional photograph. *MacNutt* aired on WDSU from 1959 through 1965. (Mary Lou Mack collection.)

This photograph captures a personal, pie-throwing appearance on *The Great MacNutt* by the Three Stooges—Curly Joe DeRita (far left), Moe Howard, and Larry Fine (right), on Wayne Mack's WDSU show. Mack was one of several TV show hosts chosen to star in the Stooges' last full-length feature film, *The Outlaws Is Coming*, released in 1965. (Mary Lou Mack collection.)

Portraying a wild and wacky movie director, Wayne Mack's Great MacNutt would spar with his young in-studio guests and introduce Three Stooges film shorts, as his fans shouted with him, "Lights! Camera! Action! Start the cotton-pickin' program!" (Paul Yacich collection.)

Wayne Mack (left) makes a guest appearance on the set with Mr. Bingle and Dr. Walrus, holiday characters who appeared every December in programs sponsored by the Maison Blanche department store. (Joe Budde collection.)

Mr. Bingle, voiced by puppeteer Oscar Isentrout, was also a television star every December on WDSU. Pictured here with one of his costars, the snowman puppet doubled as a pitchman, reminding young viewers that their favorite toys could be purchased at Maison Blanche. (Joe Budde collection.)

For the WDSU *Mr. Bingle* shows, former vaudeville entertainer and puppeteer Oscar Isentrout gave life, and the all-important voice, to Maison Blanche's snowman character, while Channel 6 staffer Al Shea was Pete the Penguin. (Joe Budde collection.)

Appearing with Mr. Bingle and his sidekick Pete the Penguin is Al Shea, the longtime WDSU staffer, who worked as a writer and producer for many early programs before making his mark as an entertainment critic on WDSU's popular *Midday*. On the Bingle shows, Shea lent his voice to the penguin character. (Al Shea collection.)

In the 1950s, Captain Sam, portrayed by Sam Page, hosted a popular WDSU show with a nautical theme. (C. F. Weber Collection, courtesy of Bergeron Studio and Gallery.)

Members of a local Girl Scout troop help fill the studio audience for Captain Sam's WDSU show. The show was one of the earliest children's programs on Channel 6, where former coworkers remember a long waiting list of children who wanted to appear on the show. (Paul Yacich collection.)

Before he hosted a popular teenage dance show, Channel 4 staff announcer John Pela starred as Captain Mercury on a short-lived but well-remembered children's show of the same name. (John Pela collection.)

In costume and on a spectacular spaceship set, Captain Mercury and his young friends capture the spirit of the space race in the early 1960s. (John Pela collection.)

Johnny Miller, in his trademark striped suit, starred as the host of *Johnny's Follies* on WJMR, which became WVUE-TV in 1965. The show's theme song was set to the tune of *Hail, Hail, the Gang's All Here*. (Carleen Graves Dunn collection.)

Three

Now the News, Sports, and Weather

If, as some say, television was little more than radio with pictures in its early years, then television news probably made even less of a statement.

The technology was primitive and the lines often blurred between content and commercials. Alec Gifford was the "Esso Reporter," reading only those news stories that had been approved by the program's sponsor, the Esso Standard Oil Company.

The city's first television news director, Bill Monroe, helped build WDSU's newsroom without any television experience under his belt. Like many of the pioneers, Monroe had tremendous experience in newspapers and radio but had no television background. Before long, his newsroom would be stocked with top-notch journalists making the most of the medium, including reporters John Corporon, Ed Planer, and Bill Slatter and photographers Mike Lala, M. J. Gauthier, Bill Delgado, and Jim Tolhurst.

One of WWL's earliest news stars, Bill Elder, would later make his name as an investigative reporter, winning a Peabody Award for his efforts. In 1975, news director Phil Johnson helped build an even stronger newsroom by hiring a young reporter named Angela Hill from Texas and pairing her on the anchor desk with Garland Robinette.

WDSU's sports staff included icons Mel Leavitt and Wayne Mack. Later WWL transformed Hap Glaudi from a newspaper star to a television sports legend. His thick New Orleans accent and offbeat style made him an unlikely anchor, but his success paved the way for a sportscaster named Buddy Diliberto on WVUE and WDSU.

For more than 50 years, Nash Roberts was the man New Orleans turned to for the lowdown on the weather. Roberts, who became known as the city's hurricane guru, worked at three local stations. He also led the way for weathercasters like Don Westbrook, the staff announcer who spent four decades delivering WWL weather forecasts.

For some local stations, their bond with the community was strengthened by the stance they took on issues affecting the city through daily editorials. At WDSU, the job of writing and delivering editorials was held by Bill Monroe, John Corporon, and Jerry Romig. In 1962, New Orleans gained another editorial voice when a gifted writer named Phil Johnson began a 37-year career as editorialist for WWL. And to think it all started with a simple "Good evening."

A former *Item* newspaper writer and associate editor, New Orleans native Bill Monroe was hired as WDSU's first news director in 1954. (C. F. Weber Collection, courtesy of Bergeron Studio and Gallery.)

After a successful career at WDSU, Bill Monroe moved on to NBC News, where he spent 10 years moderating *Meet the Press* and serving as Washington bureau chief, often appearing on *Today*. (Bill Monroe collection.)

An icon in local television news, Alec Gifford joined WDSU-TV in 1955. Besides his 10-year stint as the "Esso Reporter," he is remembered for his time as news director/anchor at WVUE and later as a WDSU anchor/reporter. (C. F. Weber Collection, courtesy of Bergeron Studio and Gallery.)

Alec Gifford delivers the news as WDSU's "Esso Reporter" in the 1950s. With the lines often blurred between broadcasters and advertisers, Esso Standard Oil Company had strict control over the content of the news program it sponsored. (C. F. Weber Collection, courtesy of Bergeron Studio and Gallery.)

Reporter John Corporon became WDSU news director in 1961. His career at the station included writing and delivering editorials and covering the Capitol beat as Washington bureau reporter. (C. F. Weber Collection, courtesy of Bergeron Studio and Gallery.)

This photograph shows the inside of the WDSU newsroom, which Bill Monroe (left) helped build as the station's first television news director. Since broadcast news was so new, the early staff consisted of mostly newspaper and radio journalists. (Author's collection.)

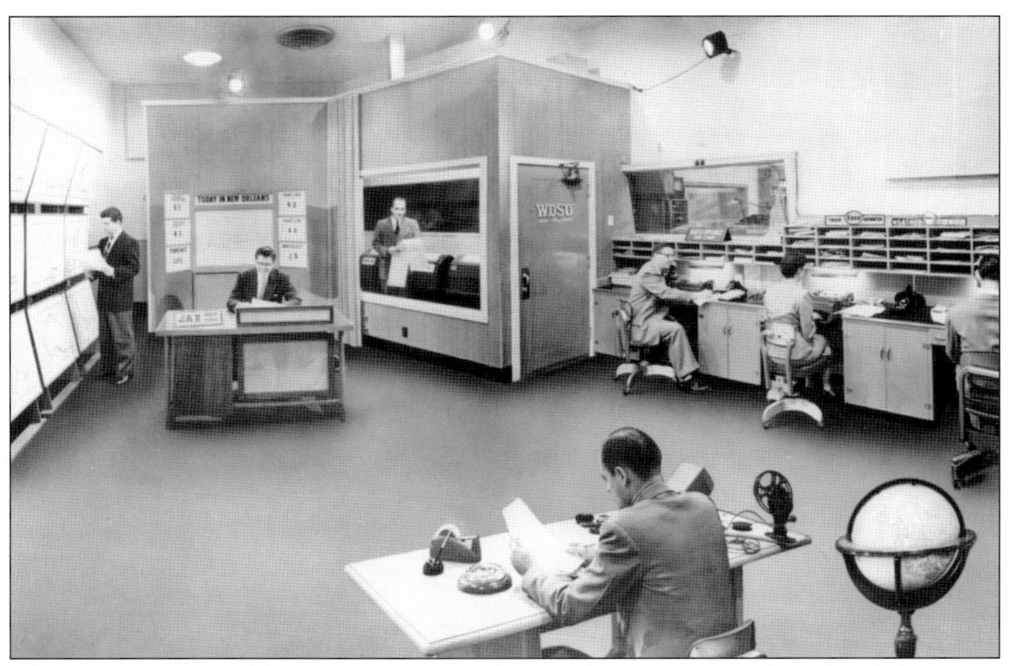

These early members of the WDSU staff at work inside the newsroom include, from left to right, Brandon Chase, at the weather map; Mel Leavitt, seated behind the *Jax World of Sports* desk; Gay Batson, inside the glass announcer's booth; announcer Tiger Flowers, seated near the door; and news director Bill Monroe (front). (Author's collection.)

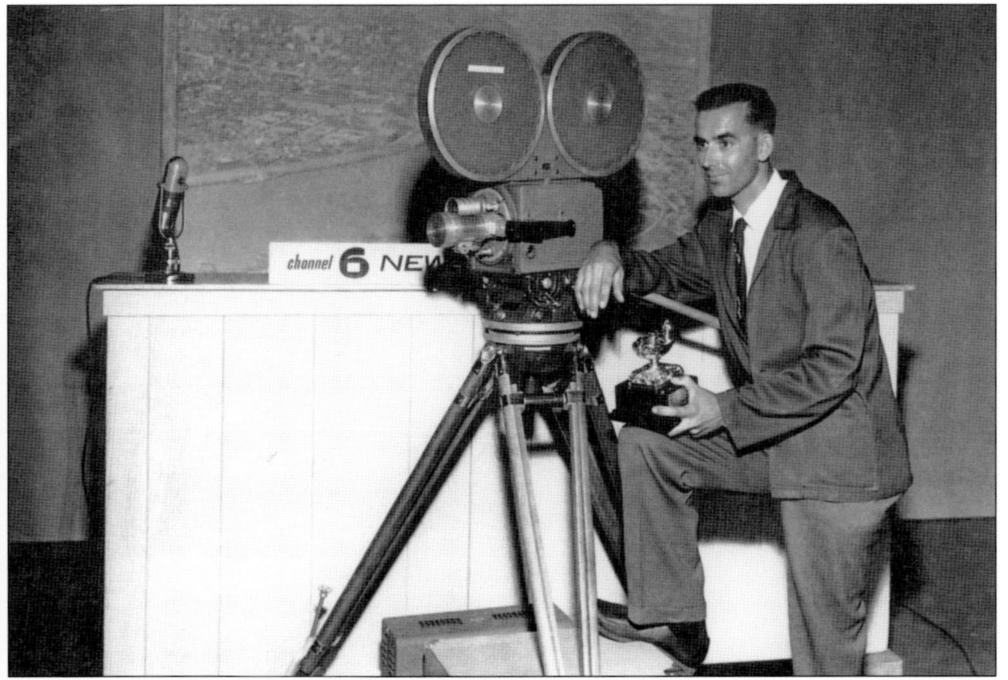

WDSU-TV photographer Mike Lala proudly displays an award won for news coverage. Lala's career spanned 30 years and included filming Lee Harvey Oswald distributing pro-Castro leaflets in New Orleans. (Paul Yacich collection.)

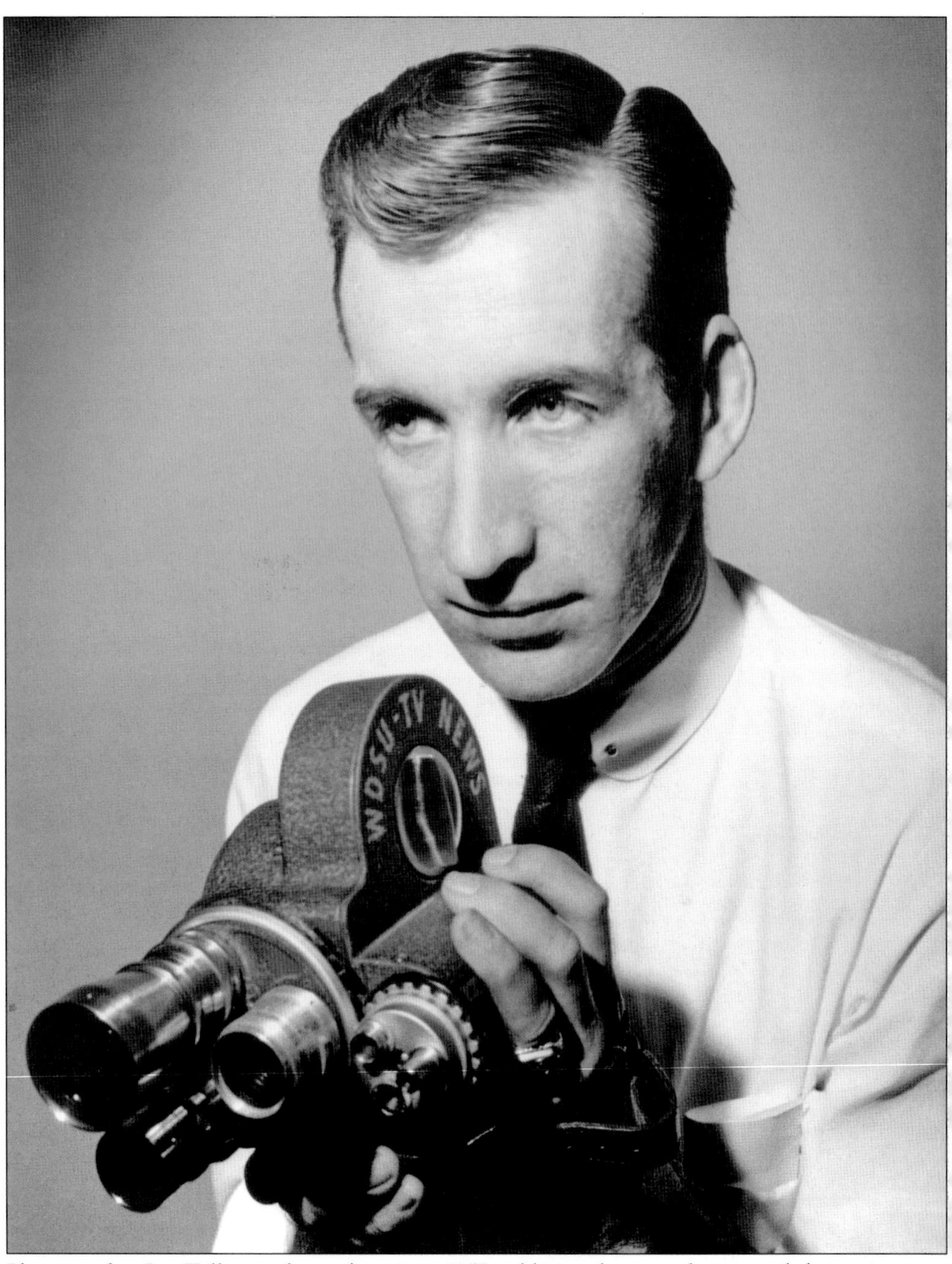

Photographer Jim Tolhurst, shown here in a 1957 publicity photograph, covered the major news stories of the 1950s and 1960s while working at WDSU. In 1966, he joined the staff of WWL-TV, specializing in documentaries. (Evann Duplantier collection.)

Photographer Jim Tolhurst, who joined the WDSU staff in 1954, documents the devastating flooding of Hurricane Betsy while on assignment for Channel 6 in 1965. (Evann Duplantier collection.)

During his career at WWL, Jim Tolhurst (right) collaborated with news director Phil Johnson to lead the team whose productions earned major national honors, including three Peabody Awards. Here they accept one for a 1972 documentary on China. (Phil Johnson collection.)

Bill Stanley, who joined the WDSU staff in 1957, hosted Channel 6's morning newscast, *The Breakfast Edition*, for three decades. Stanley began his broadcasting career in Monroe, Louisiana, before coming to work in New Orleans at WNOE radio and then WDSU. (Joe Budde collection.)

A Mississippi native, Iris Kelso began her career in print journalism with the *New Orleans States* before joining WDSU in 1967 as a political reporter. She worked for the station for 11 years before returning to her print roots as a popular columnist for the *Times-Picayune*. (Joe Budde collection.)

Some members of WDSU's 1960s stable of stars, featured in this classic promotional photograph, include, from left to right, newsmen Alec Gifford and Bill Slatter, Mel Leavitt, Terry Flettrich, editorial cartoonist John Chase, meteorologist Nash Roberts, Jan and Bob Carr, and Wayne Mack. (Bob and Jan Carr collection.)

Members of the WWL-TV news staff shown in this 1963 photograph include reporter Bob Jones (center) and news director Bill Reed (with his back to the camera). Reed came to WWL after a career in print journalism at the *Item* and *States* newspapers. (Courtesy of WWL-TV.)

This image is a behind-the-scenes view in the WWL-TV studios with the 1960s anchor team of, from left to right, Bob Jones, Hap Glaudi, and Don Westbrook. (Courtesy of WWL-TV.)

Bill Elder joined the WWL staff in 1966. A native of Opelousas, Louisiana, Elder worked in newspapers there and at a Lafayette, Louisiana, television station before coming to New Orleans. (Courtesy of WWL-TV.)

The caption for this 1960s WWL publicity photograph describes reporter Bill Elder as "aggressive and hard-hitting," two words that would be connected to his work as an anchor and reporter for more than 35 years. (Courtesy of WWL-TV.)

In addition to developing a distinctive style as anchor of WWL's noon and 5:00 p.m. newscasts for more than 30 years, Bill Elder also earned numerous awards for his investigative reporting, including the Peabody Award in 1993. (Courtesy of WWL-TV.)

The Channel 4 anchor team made up of, from left to right, Bill Elder, Hap Glaudi, and Al Duckworth is what many New Orleanians will remember when they think of television news in the early 1970s. (Courtesy of WWL-TV.)

Rosemary James went to work for WWL-TV in 1968 after a successful career at the *States-Item*. At the newspaper, she was part of the team that broke the story of Orleans Parish district attorney Jim Garrison's investigation into the assassination of Pres. John F. Kennedy. (Courtesy of WWL-TV.)

On the WWL-TV news set in the 1960s are, from left to right, weathercaster Don Westbrook, sports anchor Hap Glaudi, and newsman Jim Kincaid. Before electronic graphics superimposed the anchors' names on the screen, handmade signs did the job instead. (Courtesy of WWL-TV.)

Anchor Jim Kincaid joined WWL-TV in 1960. He left New Orleans for television jobs in St. Louis and New York. Later, as an ABC News correspondent, he covered the Vietnam War. (Courtesy of WWL-TV.)

Mel Leavitt anchors WDSU election coverage in 1956 on a set that includes signage for the sponsors. The "L" and "M" on the map represent Earl Long and Chep Morrison, candidates for governor. (Paul Yacich collection.)

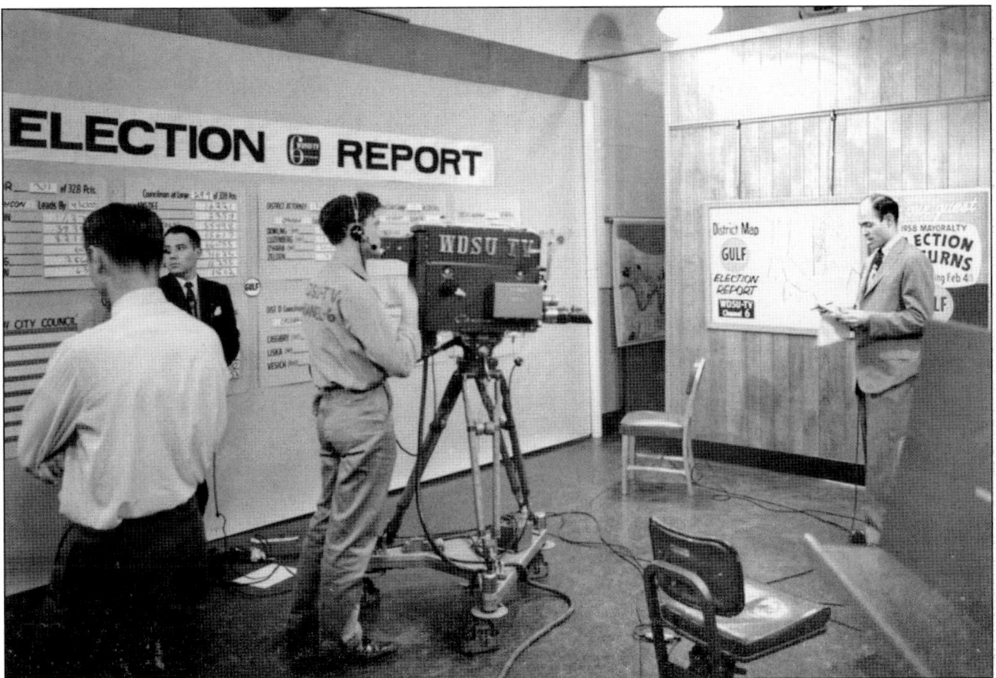

WDSU news director Bill Monroe delivers 1958 election returns inside the Channel 6 studio. To the left of the camera operator is future news director Ed Planer, then a Channel 6 reporter. (Paul Yacich collection.)

A young Alec Gifford analyzes presidential election returns in 1956, just one year after joining the WDSU news staff. Gifford's broadcasting career began with jobs at Texas radio stations before he got an offer to work at WDSU. (Paul Yacich collection.)

WDSU-TV staffers use the low-tech tools of the time to cover the results of the 1959 Louisiana governor's election. Notice the piles of paper on the floor, tossed aside whenever new election returns came in. (Paul Yacich collection.)

On election night 1964, Channel 4 tracks the local races and the presidential contest between Lyndon Johnson and Barry Goldwater. WWL's news director, Bill Reed, is standing at left, pointing out election results. (Courtesy of WWL-TV.)

New Orleans mayor Victor Schiro (left) joins future WDSU news director Doug Ramsey (right) on election night 1966, when voters considered a bond proposition for the Superdome. Updating election returns involved ripping a piece of paper off the board, tossing it on the floor and writing on the one beneath. (Joe Budde collection.)

Bill Slatter anchors WDSU's 1966 election night coverage. A fixture on Channel 6 in the 1960s, Slatter is also remembered as the anchor who interviewed Lee Harvey Oswald on WDSU, just months before the assassination of Pres. John F. Kennedy. (Joe Budde collection.)

Businessman Dave Dixon (left) joins WDSU anchor Bill Slatter (right) on the Channel 6 election set in 1966 to discuss the outcome of the domed stadium issue on the ballot that night. (Joe Budde collection.)

This is how generations of New Orleans television viewers remember weather guru Nash Roberts: in front of his map, with a felt-tip marker, calmly and accurately delivering his weather forecast. In the 1950s and 1960s, Roberts's WDSU weathercasts were sponsored by Jax Beer, among other businesses. (C. F. Weber Collection, courtesy of Bergeron Studio and Gallery.)

Nash Roberts began his career at WDSU, as the first television meteorologist in the South, when he was hired on a freelance basis to help track a hurricane. That was a role he would build his career on, working for three local stations over five decades. (C. F. Weber Collection, courtesy of Bergeron Studio and Gallery.)

A scientist first and television star second, Nash Roberts was the first meteorologist to plot a typhoon by flying into one during World War II. He came back to New Orleans after the war and opened a weather consulting service before television came calling. (C. F. Weber Collection, courtesy of Bergeron Studio and Gallery.)

Here is a promotional portrait of Nash Roberts during his heyday at WDSU. Later, during his stints at WVUE and WWL, whenever a hurricane threatened, the question "What does Nash say?" was like a mantra for New Orleanians gauging the path of an approaching storm. (Author's collection.)

Originally hired as a WWL staff announcer in 1960, Don Westbrook later added weathercaster to his job title. He would continue in that role for more than 35 years. (Don Westbrook collection.)

In the earliest days of television, the line between news and commercials was often crossed, as is the case on the set of this early WWL-TV weathercast with Don Westbrook. (Don Westbrook collection.)

Described as the "consummate communicator," Don Westbrook became a fixture on WWL-TV's morning and noon newscasts, in addition to continuing in his original role as a station announcer. (Courtesy of WWL-TV.)

Bart Darby, who spent 20 years at WDSU, came to New Orleans in 1962 from Birmingham, Alabama. In New Orleans, Darby is remembered as a weathercaster, host, and one of the announcers for the annual Carnival broadcast of the Meeting of the Courts of Rex and Comus. (Joe Budde collection.)

Al Duckworth joined the WWL-TV weather staff in 1968, coming to New Orleans after working at stations in Oklahoma, Ohio, and Florida. His career also included positions at WDSU and WVUE. (Courtesy of WWL-TV.)

Al Duckworth poses for a publicity photograph in the WWL Weather Center. In its promotions, the station boasted that Duckworth's "expertise is complemented by a weather center equipped with the most recent scientific weather detection devices." (Courtesy of WWL-TV.)

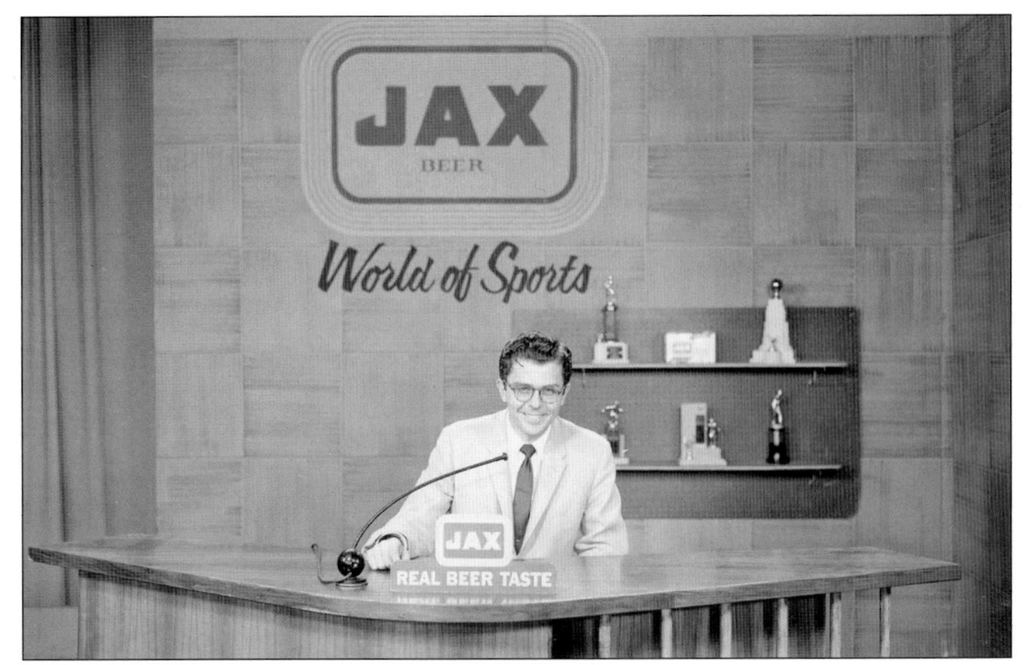

Mel Leavitt held many on-air roles at WDSU and four other local stations during his career, but he originally came to Channel 6 in 1949 to work as a sports announcer. Here he hosts *Jax World of Sports*, with prominent mention of its sponsor, the Jackson Brewing Company. (C. F. Weber Collection, courtesy of Bergeron Studio and Gallery.)

Wayne Mack came to WDSU as a sports announcer and host in 1958 after working at radio and television stations across the country. He was named Channel 6 sports director in 1966 and is also remembered as the radio voice of the New Orleans Saints in the 1970s. (Paul Yacich collection.)

Hap Glaudi changed the face of local television when he joined WWL in 1961. Besides giving sports news and scores, Glaudi offered commentaries and birthday and anniversary greetings and chatted freely with his coanchors, something that was not common at the time. (Courtesy of WWL-TV.)

WWL sports director Hap Glaudi (right) interviews boxing legend Joe Louis inside the ring in 1962. (Courtesy of WWL-TV.)

Newspaperman Hap Glaudi, who built a following as a sports editor and columnist for the *Item*, was an unlikely choice for WWL's sports director. The Jesuit High School graduate with a distinctly New Orleans accent went on to become a local broadcasting legend. (Courtesy of WWL-TV.)

Sports director Hap Glaudi welcomes heavyweight boxing champion Rocky Marciano (above, right) to the WWL sports desk in the 1960s. (Courtesy of WWL-TV.)

Jim Henderson was given the unenviable task of replacing Hap Glaudi as WWL's sports anchor when he joined the station in 1978. Irate fans picketed the station when Henderson arrived, but the former English teacher's superb writing and broadcasting skills soon won over fans. (Jim Henderson collection.)

Two giants in local sports broadcasting, Jim Henderson (left) and Buddy Diliberto (right), pose for a picture promoting their support for WYES-TV's Showboat Auction. A New Orleans native, "Buddy D." was known as much for his malapropisms as for his frank criticism of the New Orleans Saints during his time at WVUE, WDSU, and WWL radio. (Jim Henderson collection.)

Say what you will about their fashions, but these WWL stars were the ones people watched in the 1970s. The members of the morning and weekend anchor teams are, from left to right, news anchor Eric Paulsen, weathercaster Don Westbrook, meteorologist Al Duckworth, sports anchor Bob Krieger, and news anchor Dennis Woltering. (Courtesy of WWL-TV.)

Garland Robinette, a native of Boutte, Louisiana, joined Channel 4 in 1972. Besides his work as an anchor, Robinette also earned respect as an environmental reporter during his 18-year career with the station. (Courtesy of WWL-TV.)

Angela Hill came to New Orleans in April 1975 from Corpus Christi, Texas, for a job at WWL-TV as a consumer reporter. (Courtesy of WWL-TV.)

Within months of her arrival at WWL, Angela Hill was coanchoring the station's top-rated 6:00 p.m. and 10:00 p.m. newscasts. She would also go on to host her own successful daily talk show in the 1980s and 1990s. (Courtesy of WWL-TV.)

On the set of a WWL newscast in the 1970s are news anchors Garland Robinette (left) and Angela Hill, not long after they joined the station. Joining them at the anchor desk are sports director Hap Glaudi and meteorologist Al Duckworth (standing). (Courtesy of WWL-TV.)

This is a promotional photograph of the enormously popular WWL-TV anchor team in the 1970s, which included, from left to right, news anchors Garland Robinette and Angela Hill, sports director Hap Glaudi, and meteorologist Al Duckworth. (Courtesy of WWL-TV.)

First paired as anchors in 1975, Angela Hill and Garland Robinette, the city's longest-running television anchor team, helped propel WWL's *Eyewitness News* into a ratings juggernaut in the 1970s and 1980s. Because of their enormous popularity, the couple's marriage in 1978 caught the city by storm. (Courtesy of WWL-TV.)

WWL's position of dominance in the local TV ratings was solidified in the 1970s and 1980s by these three superstars, with Bill Elder (left) anchoring at noon and 5:00 p.m. and Angela Hill and Garland Robinette at 6:00 p.m. and 10:00 p.m. (Courtesy of WWL-TV.)

A New Orleans native whose journalism career began at the *Item* newspaper, Phil Johnson went to work at WWL-TV as promotion director in 1960. During his long career at the station, Johnson would earn accolades for his work as news director, documentary producer, editorialist, and assistant general manager. (Phil Johnson collection.)

Through his nightly editorials, Phil Johnson, familiar for his bearded face and trademark phrase, "Good evening," helped fulfill a mission of public service set for WWL-TV by its Jesuit owners at Loyola University and the station's general manager, J. Michael Early. (Courtesy of WWL-TV.)

Phil Johnson is shown here without his familiar beard, shortly after delivering his first WWL editorial on March 26, 1962. That first editorial explained, "Beginning today and every weekday hereafter, this station will present editorial opinion, a living vigorous commentary on all things pertaining to New Orleans, its people and its future." (Phil Johnson collection.)

Phil Johnson, shown here on the Channel 4 set in the 1960s, delivered WWL-TV editorials from 1962 until 1999, making his the longest tenure of any broadcast editorialist in the country. (Courtesy of WWL-TV.)

A 1960s innovation at WDSU involved someone who had been expressing his ideas for years in newspapers: editorial cartoonist John Churchill Chase. His television cartoons, often featuring this "Little Man" character introducing his segments, appeared nightly at 6:00 p.m. and 10:00 p.m. (Courtesy of WDSU-TV.)

This was John Chase's first editorial cartoon for television, premiering on WDSU on August 17, 1964. It showed Chase's familiar "Little Man" peering out of the screen at "The People" he represented. Chase's successful WDSU career followed stints as the cartoonist for the *Item*, *Times-Picayune*, *States*, and *States-Item* newspapers. (Joe Budde collection.)

Four

MIDDAY MEMORIES

"It's high noon in the Crescent City and time to catch up on the news with a lot of people on *Midday*—30 minutes of information and fun." Every day at noon, those words would open one of the most enduring and popular programs in New Orleans television history.

From the 1950s through the 1970s, Terry Flettrich was host, producer, and driving force behind the program, leading a cast that included Wayne Mack, Al Shea, Nash Roberts, Bob and Jan Carr, and WDSU newscasters Alec Gifford, Iris Kelso, and Ed Planer. Music from Pete Laudeman and segments in the kitchen helped entertain the in-studio audience, made up of local women's groups, all dressed up for their television appearance.

A WDSU promotional brochure explained some regular segments: "One day you'll hear a novelist and a sculptor interviewed. On another day, a man from the Better Business Bureau warns women of the wily. This is New Orleans' most popular, highly-rated women's program."

Announcer and cohost Wayne Mack referred to himself as the show's second banana, and the title fit. Though he was known as both a sportscaster and kid show host, on *Midday*, Mack's witty and charming personality helped entertain the ladies who made the trek down to the French Quarter to appear on the show.

Entertainment critic Al Shea began his career behind the scenes on *Midday* and other WDSU programs, but he ended up in front of the camera, as entertainment critic and celebrity interviewer.

On Shea's recommendation, Channel 6's married couple, Bob and Jan Carr, began their local TV careers with a morning show, *Second Cup*, but became well known for their family-friendly features on *Midday*. Their familiar greeting—"This is Bob. And this is Jan."—still brings a smile to the faces of the people who grew up watching them.

Throw in the news headlines from Alec Gifford, an interview with the mayor by Iris Kelso, and the weather from Nash Roberts, and one can see why *Midday* is often referred to as local programming at its finest. It set the standard for a television genre in New Orleans—much more than just "30 minutes of information and fun."

Terry Flettrich poses for a photograph on the set of *Midday*, which she produced and hosted for more than 20 years. As a WDSU promotional piece explained, "Terry surrounds herself with the cream of WDSU's scintillating talent, who catch her spark: Alec Gifford, top-rated news analyst; Bob and Jan Carr, representing the parentally youthful; Wayne Mack for pixie humor; Nash Roberts, meteorologist." (C. F. Weber Collection, courtesy of Bergeron Studio and Gallery.)

Members of local civic and social groups made up the studio audience for *Midday*, where a ladies' hat was clearly de rigueur. (C. F. Weber Collection, courtesy of Bergeron Studio and Gallery.)

One of the many *Midday* roles for Wayne Mack (standing at left) was to entertain and interview, in his trademark playful way, the ladies of the studio audience. "Every day it was my job to bring them in and give them sandwiches and warm Pepsis and Angelo Brocato cookies," Mack once recalled. (Al Shea collection.)

This *Midday* promotional piece illustrates six of the many reasons the program was so successful. Clockwise from top left are Jan Carr, Bob Carr, Terry Flettrich, Al Shea, newsman Ed Planer, Wayne Mack, and Nash Roberts. (Author's collection.)

Vera Massey preceded Terry Flettrich as *Midday* hostess on the program's predecessor, *Our House*. Most viewers did not know that Massey suffered from a debilitating illness and had to be carried on and off of the set. When she became too ill to continue as host, Flettrich was tapped to take over her duties. (Paul Yacich collection.)

A classic 1960s WDSU promotional piece for host Terry Flettrich describes, "Terry is a point of view. She believes that a woman has a mind and wants to use it, wants to find out more about a world that's whirling her around so fast she hardly can find time to hang on." (Terry Flettrich Rohe collection.)

WDSU once billed *Midday* host Terry Flettrich as "another living reason why people look to WDSU-TV. They know that Channel 6 personalities are the most vital, interesting people in their fields and have earned the right to a viewer's attention." (Terry Flettrich Rohe collection.)

Terry Flettrich (left) and Alec Gifford (right) welcome New Orleans mayor Victor Schiro to the *Midday* news desk. As the station promoted those interviews: "Once each week, the Mayor of New Orleans answers questions that probe, pursue and enlighten." (C. F. Weber Collection, courtesy of Bergeron Studio and Gallery.)

Al Shea (right) interviews singer Brenda Lee on the *Midday* set in 1967. Shea was with *Midday* from the beginning, though not always in front of the camera. Originally a commercial writer for the program, he later earned a spot as entertainment editor. (Al Shea collection.)

Midday would send Al Shea (right) to Broadway and Hollywood each year to interview celebrities like Dick Van Dyke. Here the former Channel 6 star talks to Shea about his 1968 film, *Chitty Chitty Bang Bang.* (Al Shea collection.)

Al Shea (left) interviews actor Burt Reynolds during one of his *Midday* trips to the West Coast, stargazing for the audience at home. (Al Shea collection.)

Al Shea is pictured on the job as entertainment critic for *Midday*. After a long career in local broadcasting and print journalism, Shea continues on the entertainment beat for WYES-TV's weekly arts and entertainment program *Steppin' Out*. (Al Shea collection.)

Movie and nightclub performer Jayne Mansfield (center) looms large in *Midday* history. In 1967, Mansfield was traveling from the Mississippi Gulf Coast to New Orleans for a *Midday* interview with Al Shea when she was killed in an automobile accident. Here Mansfield and her children sit for an interview with Bob Carr (right), filling in as host of Mel Leavitt's *Byline*. (Paul Yacich collection.)

This WDSU advertisement for Bob and Jan Carr properly promotes them as "a happy marriage of entertainment and good common sense." As the parents of four children, the Carrs' *Midday* segments focused mostly on family issues but also included fashion and lifestyle pieces. (Bob and Jan Carr collection.)

Bob and Jan Carr began their New Orleans broadcasting careers in 1960 at WWL radio and later auditioned for a job at WDSU-TV at the urging of their friend Al Shea. They soon began hosting their own daily morning show on Channel 6, *Second Cup*, in 1961 before joining the *Midday* crew in 1963. (Bob and Jan Carr collection.)

Bob and Jan Carr's popular morning program, *Second Cup*, was broadcast from the roof of the newly constructed Royal Orleans Hotel in the French Quarter. As a promotional brochure described, "The fresh and personable young couple are hosts for this fast-moving half-hour of entertainment and information." (Bob and Jan Carr collection.)

Hosted by this handsome young married couple and offering a spectacular view of the city from atop the Royal Orleans Hotel, *Second Cup* "is easily the best 'looking' show in town," boasted a 1960s WDSU promotional piece. (Bob and Jan Carr collection.)

Midday cohost Wayne Mack, in costume for a Mardi Gras show, mugs for the camera. (Al Shea collection.)

In costume for the Mardi Gras *Midday* show are cast members, from left to right, Wayne Mack, Al Shea, Jan Carr, and Terry Flettrich. (Al Shea collection.)

Al Shea joined the WDSU staff in 1955 as a floor director. He later became a writer and producer for *Midday* and *Second Cup*, where he began presenting entertainment reviews and celebrity interviews. (Al Shea collection.)

As entertainment editor, Al Shea's theater and film reviews became his trademark on *Midday* for the better part of a decade, from 1963 to 1973. (Al Shea collection.)

Chef Paul Blange (left) poses with *Midday* host Terry Flettrich. Chef Blange, besides appearing in numerous cooking segments on *Midday*, was the head chef at Brennan's, where he invented many of the restaurant's legendary dishes, including bananas Foster. (Joe Budde collection.)

Wayne Mack (left) interviews actor Van Johnson, surrounded by the ladies of the *Midday* studio audience. The program regularly featured interviews with visiting celebrities appearing at the Roosevelt Hotel's Blue Room and other venues. (Joe Budde collection.)

Wayne Mack (right) welcomes ventriloquist Edgar Bergen and his ever-present sidekick Charlie McCarthy to the *Midday* interview set in the 1960s. (Mary Lou Mack collection.)

WDSU program director Jerry Romig (right) accepts an award from New Orleans mayor Victor Schiro on the *Midday* set. Romig's television career began in 1955, when he joined the WDSU staff. Later, as producer, editorialist, program director and station vice president, he was instrumental in the success of many WDSU productions, including *Midday* and *Second Cup*. (Joe Budde collection.)

Five

THE ORIGINAL MUST-SEE TV

They were the shows that every teenager in town dropped what they were doing on Saturdays to watch. *Morgus the Magnificent* featured the madcap scientist whose experiments always backfired. John Pela hosted the teenage dance party that became a television institution. Both embody the originality and personality of local shows during the golden age of New Orleans television.

Dr. Momus Alexander Morgus first appeared on television screens in 1959, becoming an overnight sensation for WWL. Each week, with help from a talking skull named Eric and a hulking assistant named Chopsley, Morgus updated viewers on his scientific exploits in between segments of horror movies.

Known as *Saturday Hop* when it premiered, Channel 4's dance show was renamed for WWL staff announcer John Pela after he took over as host in 1961. Local teenagers got their chance at stardom, whether invited to be part of the group of teens who danced on the show or as a "specialty" dancer from the Hazel Romano or Tony Bevinetto dancing schools. The songs, the clothes, the dances—it was all vintage 1960s and a huge part of many young lives. *Prep Quiz Bowl* also made an impact on local teenagers in a more studious way. The WDSU program, hosted by Mel Leavitt, pitted student teams from local high schools against each other. Questions tested their knowledge of everything from classical music and fine art to science and geography.

For years, Leavitt was also known as "Mr. Mardi Gras" for his role in Carnival coverage, narrating the nightly broadcasts of the parades from atop WDSU's Royal Street balcony. Channel 6's coverage also included the annual broadcast of the Meeting of the Courts of Rex and Comus, with narrator Gay Batson's distinctive voice helping to close out Carnival for decades.

Jim Metcalf makes any list of distinguished voices, first as a WWL newsman, then as host of the immensely popular *A Sunday Journal*. The Sunday night program featured essays and poems written by Metcalf, often set to music. Metcalf began every *Journal* by intoning "Please to begin." What better way is there to open our look back at some beloved programs and personalities?

Here is a rare, behind-the-scenes look inside the old city icehouse laboratory of Dr. Momus Alexander Morgus. His Channel 4 Saturday night program premiered as *The House of Shock* on January 3, 1959, and became an instant hit for the station, which had signed on just two years earlier. (Tom Bourgeois collection.)

This is either a promotional photograph for a sponsoring soft drink or a scene from a Morgus experiment. Perhaps not so coincidentally, the WWL studios where Dr. Morgus first broadcast his experiments were housed in a converted 7-UP bottling plant on North Rampart Street. (Courtesy of WWL-TV.)

This view is a look inside the lab at WWL-TV where Morgus first began televising his experiments. The master was always assisted by his devoted but clumsy henchman, Chopsley, and Eric, the former lab assistant whose brain was preserved as a talking skull. (Author's collection.)

Morgus the Magnificent, with help from Eric and Chopsley (originally played by Tommy George) presented televised experiments on four different New Orleans stations over the past 50 years. For a time, Morgus also hosted a daily weather show, which featured him wringing out a "humidity rag" as part of his forecasts. (Author's collection.)

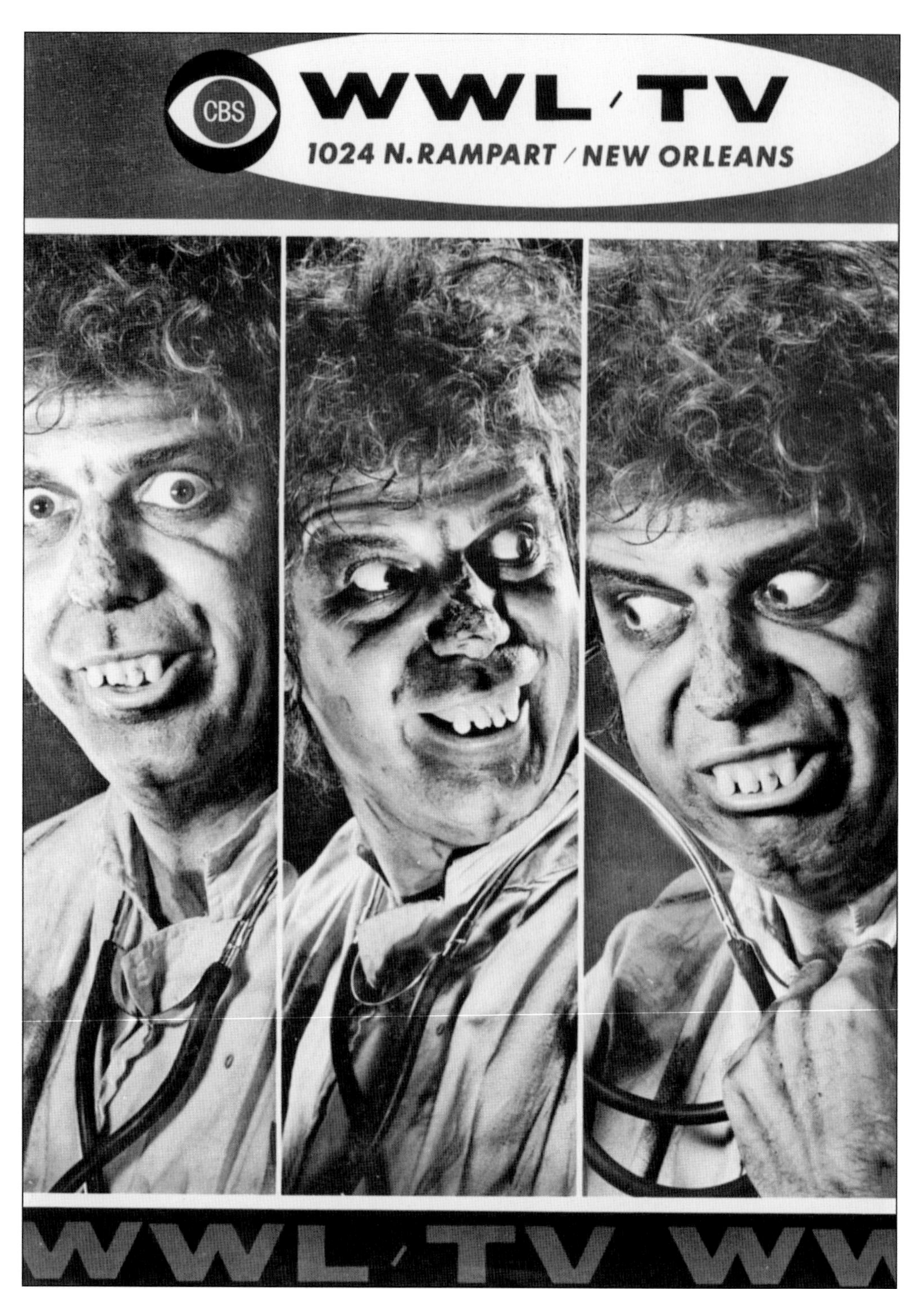

A promotional piece for one of the most popular WWL programs of the 1960s features the three faces of Morgus. His experiments aimed to carry out the mission of the Higher Order, a secret society dedicated to preserving and elevating the planet through science. Unfortunately, Morgus's experiments always seemed to backfire on him by the end of each episode. (Author's collection.)

With a dozen or so teenaged couples slow-dancing behind him, John Pela poses for a picture on the set of *Saturday Hop*, WWL's popular Saturday afternoon teen dance show, which was later renamed *The John Pela Show*. (Courtesy of WWL-TV.)

Everything about *The John Pela Show* was vintage 1960s, including its logo, which was superimposed over the silhouettes of the dancers at the start of this program. (Courtesy of WWL-TV.)

Announcer John Pela's distinctive voice appeared on commercials and station identification spots throughout his long career at WWL-TV. In 1961, approximately two weeks into his tenure at Channel 4, he was tapped to take over the station's Saturday afternoon dance show. (Courtesy of WWL-TV.)

The John Pela Show featured a changing backdrop every week, like this steamboat set. The sets, designed and built by WWL production designer Juozas Bakshis and his team, reflected holidays, seasons of the year, and other themes. (John Pela collection.)

The dancers and music were the main attractions, but John Pela also kept his teen guests entertained with games like Pela's Past, in which contestants were given clues about a song played on a past show, then asked to identify the artist and the song. (John Pela collection.)

In addition to teenagers who earned a guest shot on *The John Pela Show*, there were regular attractions, including performers from the Hazel Romano and Tony Bevinetto dancing schools. Here some of the Romano dancers pose for a picture with the show's host. (John Pela collection.)

Though not the original host of WWL's *Saturday Hop*, John Pela is certainly the best-known. His Saturday afternoon dance show, which ran from 1961 to 1972, was a local version of *American Bandstand* but twice as popular in New Orleans. (John Pela collection.)

Mel Leavitt (right) is fondly remembered as host of *Prep Quiz Bowl*, the local high school trivia challenge that premiered on WDSU in 1968, originally sponsored by the Pipe Council, a local trade association. (Joe Budde collection.)

Host Mel Leavitt poses for a photograph with *Prep Quiz Bowl* producer Margie Larson. The long-running program pitted two four-student teams from area high schools in a competition designed to test their general knowledge and quick recall. (Joe Budde collection.)

The program had a new name, *Varsity Quiz Bowl*, but a familiar host when it moved from WDSU to public television station WYES in 1972. Hosted by Mel Leavitt, and later Dan Milham, the program continued to air there well into the next decade. (Courtesy of WYES-TV.)

Jim Metcalf's successful Sunday night program A *Sunday Journal*, which featured Metcalf's essays and vignettes, often set to music, earned WWL a coveted Peabody Award in 1975. (Courtesy of WWL-TV.)

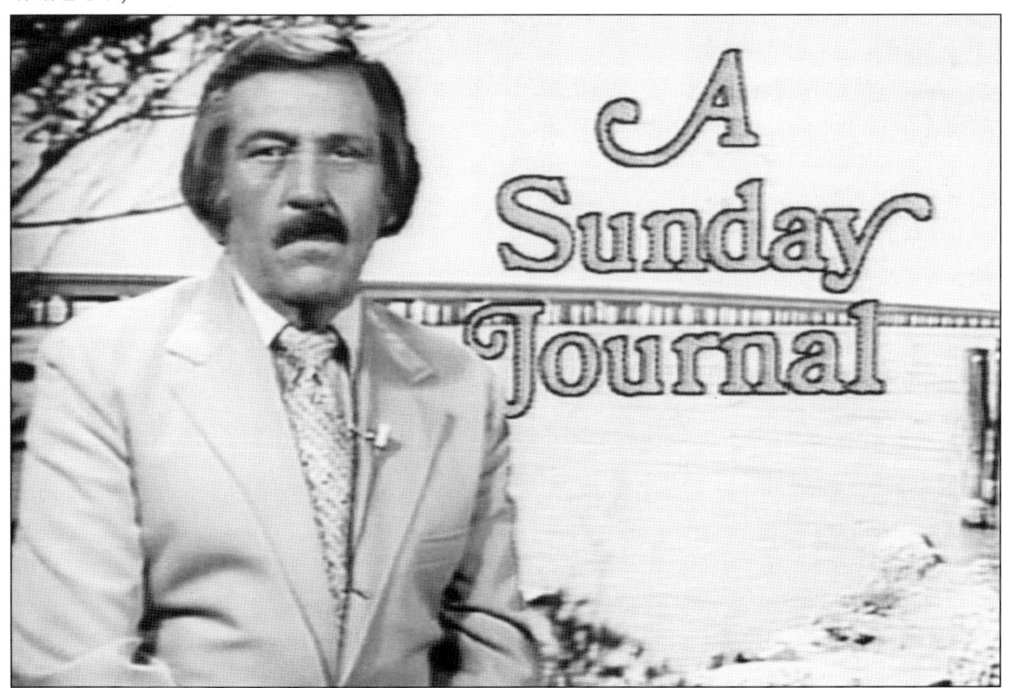

In awarding *Sunday Journal* a Peabody Award, the judges said Jim Metcalf "blended each facet of the program together with skill and aplomb," adding, "His efforts were marked by good writing, excellent photography, and artistic presentation." (Courtesy of WWL-TV.)

Originally from Texas, Jim Metcalf spent a decade covering news for Channel 4 before carving out a niche as the popular host of *A Sunday Journal* in the 1970s. (Courtesy of WWL-TV.)

WWL-TV's Sunday night program *Shades of New Orleans* introduced viewers to the wonders of color television, with color vignettes set to music. During its 12-year run, the program was produced at various times by WWL staffers Leo Willette, Marilyn Sublette Imwalle, Mickey Wellman, and Patricia Gormin. (Courtesy of WWL-TV.)

WWL staff announcer Leo Willette, who worked at the station from 1961 to 1965, was one of the early hosts of *Shades of New Orleans*. He often boasted that his was the first face that viewers saw when Channel 4 began using color television cameras on *Shades*. (Linda Cuthbert collection.)

WDSU announcer Bart Darby hosted a 1960s daytime favorite on Channel 6, *Dialing for Dollars*, the local version of a nationally syndicated audience participation game show. (Joe Budde collection.)

From 1949 to 1972, WDSU coverage of Carnival parades in the French Quarter meant hauling cameras both out of the front door and onto the station balcony, which overlooked Royal Street. The setting offered spectacular views of the floats weaving their way through the narrow streets, illuminated by flambeaux. (C. F. Weber Collection, courtesy of Bergeron Studio and Gallery.)

Mel Leavitt narrates WDSU coverage of a Carnival parades from atop the Channel 6 balcony on Royal Street. Many New Orleanians have memories of watching the parades along the route, then rushing home to watch the festivities again on television. (C. F. Weber Collection, courtesy of Bergeron Studio and Gallery.)

Mel Leavitt broadcast more than 150 parades during his years with WDSU, often collaborating with Channel 6 stalwart Paul Yacich, who produced and directed decades of Carnival coverage. (Arthur Hardy collection.)

The float of a Mardi Gras monarch passes by the WDSU balcony during a Carnival broadcast. For this particular parade, commentator Mel Leavitt (center) is joined on the balcony by music legend Pete Fountain (right) and his family. (Joe Budde collection.)

Mel Leavitt, shown here at the WDSU broadcasting booth on Canal Street, cherished his role as Carnival commentator, covering Mardi Gras parades for well over two decades on WDSU. (Arthur Hardy collection.)

WDSU uses the latest in mobile broadcasting technology, by hoisting its cameras and transmitters atop a converted bus, to broadcast Mardi Gras festivities on Canal Street. (C. F. Weber Collection, courtesy of Bergeron Studio and Gallery.)

WDSU's mobile broadcasting unit covers Carnival from in front of the D. H. Holmes Canal Street department store (right), which donned banners honoring the Fat Tuesday parades of Comus and Rex. (C. F. Weber Collection, courtesy of Bergeron Studio and Gallery.)

Canal Street was Carnival central for WDSU throughout the 1950s and 1960s, with cameras set up to cover the parades as they rolled downtown. As the city's first television station, WDSU was the first to broadcast live Carnival coverage in 1949. (Joe Budde collection.)

WDSU anchor Terry Flettrich playfully poses with a Mardi Gras masker on the station's Canal Street reviewing stand. At left is anchor Mel Leavitt, dressed as a cowboy. (Joe Budde collection.)

Members of the WDSU production crew donned formal wear for their work on the broadcast that marked the climax of Carnival each year, the Meeting of the Courts of Rex and Comus. The first WDSU broadcast of the event in 1953 was hosted by announcer Tiger Flowers. (Daphne Muller Eyrich collection.)

WDSU chief announcer Gay Batson will forever be linked to the Meeting of the Courts broadcasts, which he narrated for more than two decades. Actress Ann Meric joined Batson as cohost in the late 1960s and continued in the role with Bart Darby and Terry Gerstner after Batson's death in 1976. (C. F. Weber Collection, courtesy of Bergeron Studio and Gallery.)

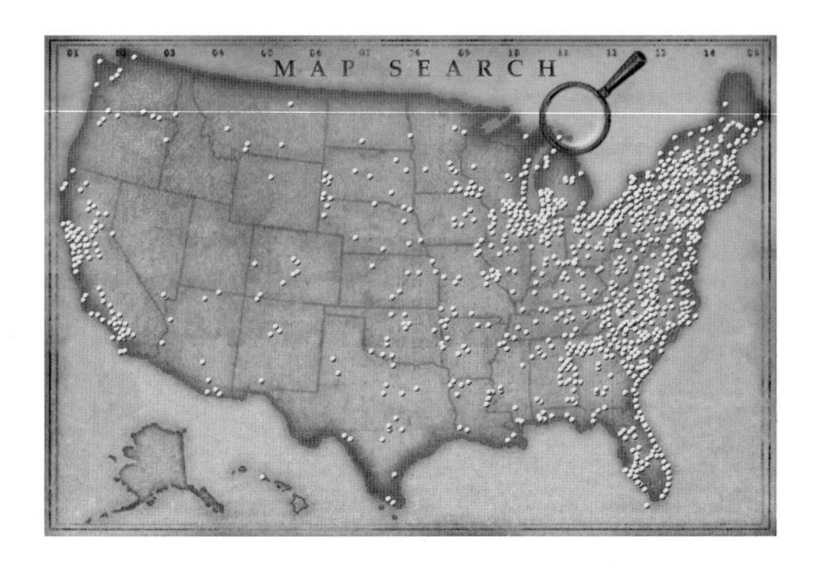

MAP SEARCH